FARGO PUBLIC LIBRARY

On the Move
TRAINS

Ursula Pang

PowerKiDS press

PK Beginners

Here comes the train!

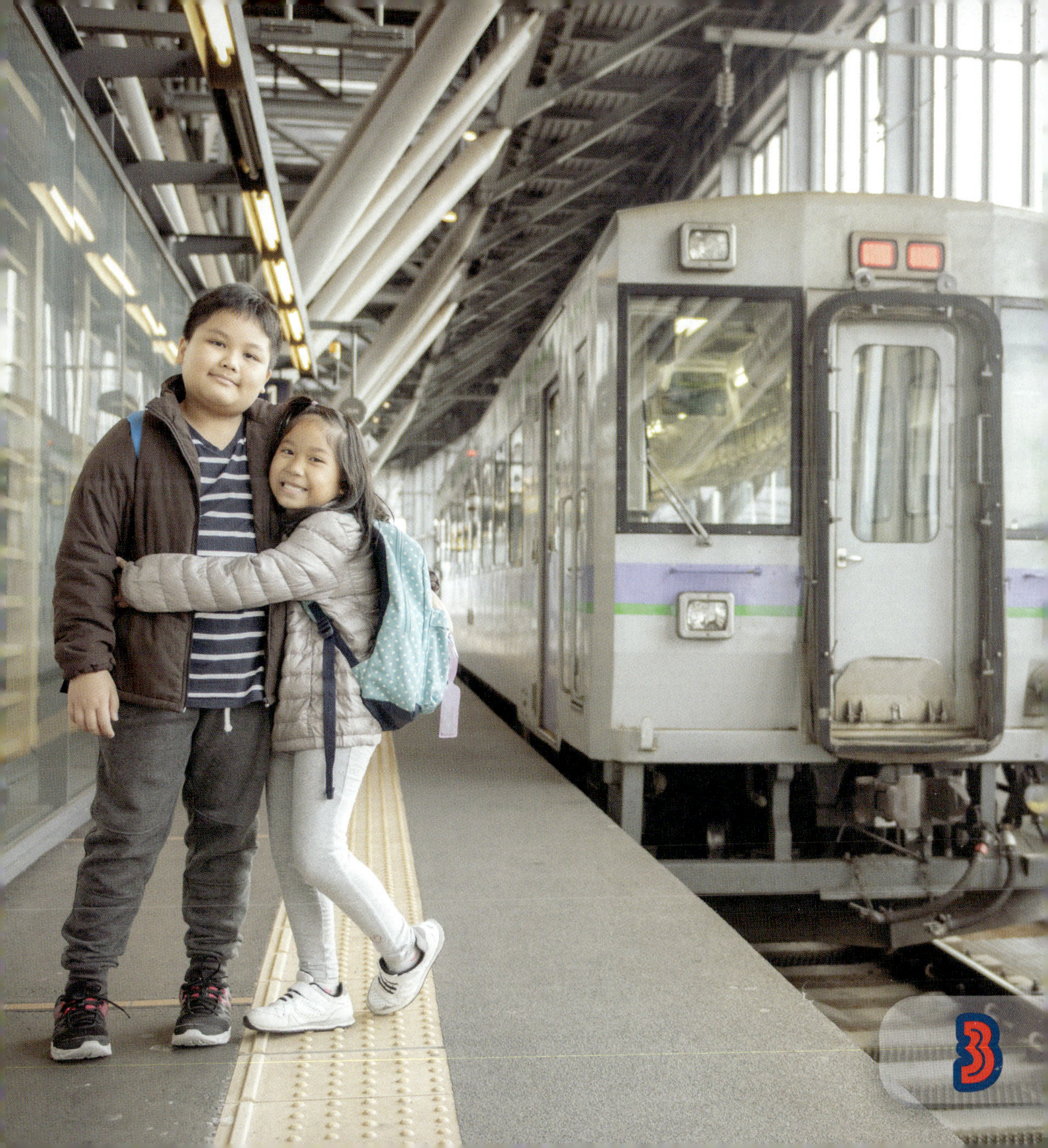

Be safe around trains.
Trains are big and fast!

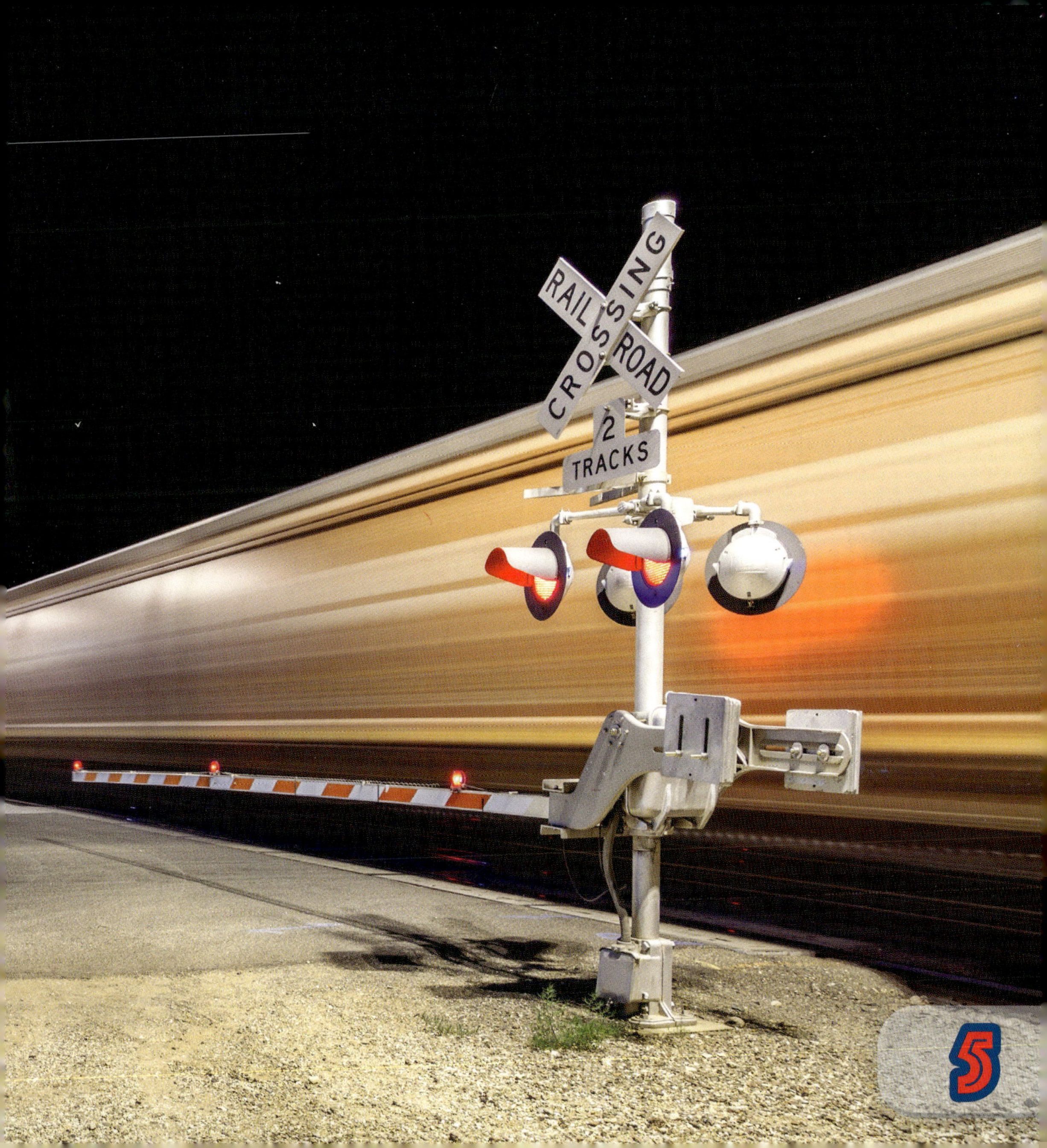
RAIL ROAD
CROSSING
2
TRACKS

Trains can have many cars.

CAR

Train cars have many wheels.
The wheels move on railroad tracks.

WHEEL

RAILROAD TRACK

The first car
is the engine.
The engine pulls the
other cars.

ENGINE ↓

11

A train driver is called an engineer.

ENGINEER

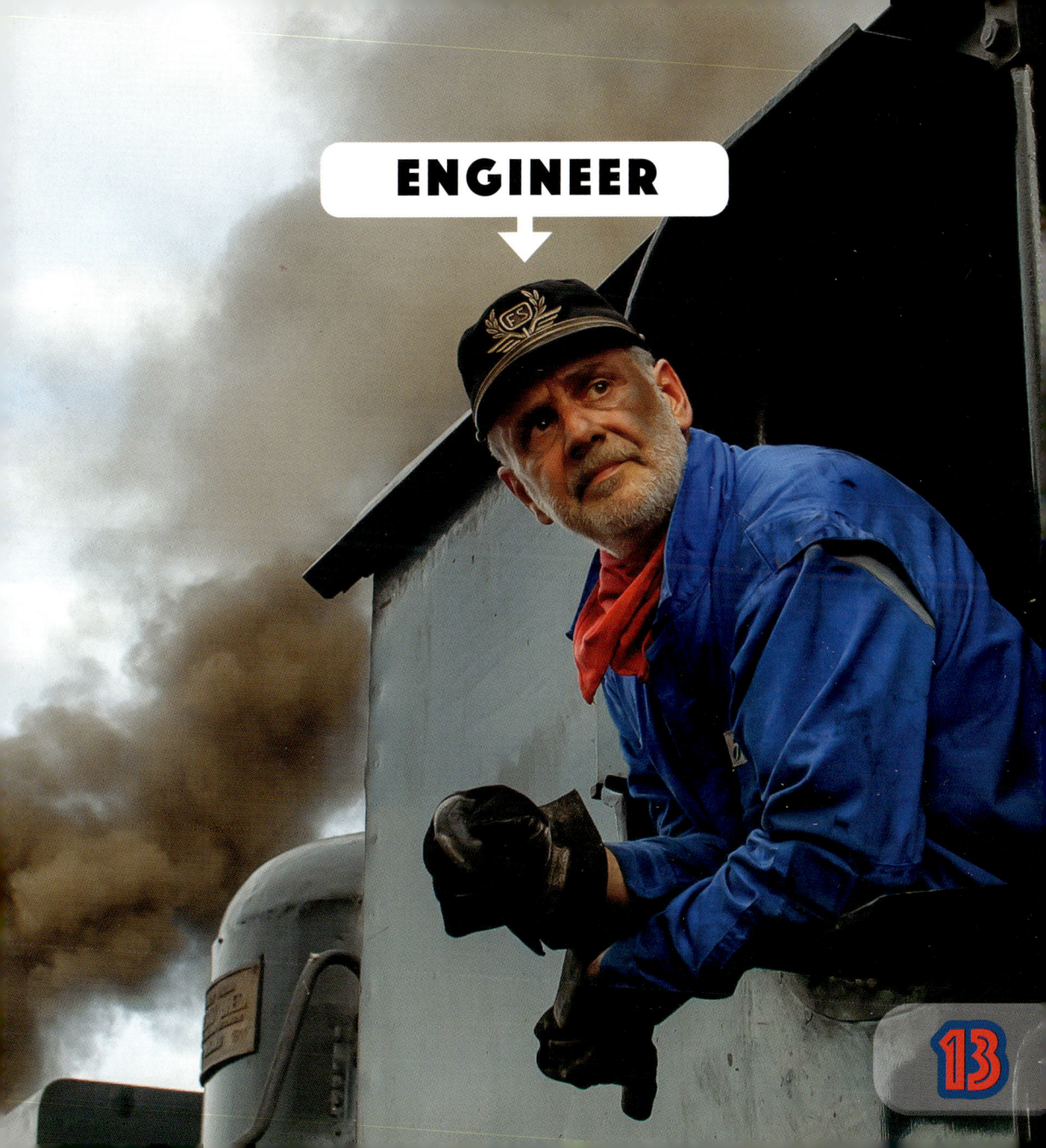

13

Some cars have seats for people. It's safe to stay in your seat.

15

A subway is
an underground train!

17

Some trains carry goods. The goods are called cargo.

CARGO ↓

19

Some countries have bullet trains.
They are very fast!

21

Have you ever taken a trip on a train?

23

Published in 2025 by The Rosen Publishing Group, Inc.
2544 Clinton Street, Buffalo, NY 14224

Copyright © 2025 by The Rosen Publishing Group, Inc.

All rights reserved. No part of this book may be reproduced in any form without permission in writing from the publisher, except by a reviewer.

First Edition

Editor: Greg Roza
Book Design: Rachel Rising

Photo Credits: Cover, p. 1 Boxcar Media/Shutterstock.com; p. 3 kwanchai.c/Shutterstock.com; p. 5 travelview/Shutterstock.com; p. 7 Guitar photographer/Shutterstock.com; p. 9 Raman Venin/Shutterstock.com; p. 11 lihaoming1234567/Shutterstock.com; p. 13 esherez/Shutterstock.com; p. 15 talyonen/Shutterstock.com; p. 17 Marek Masik/Shutterstock.com; p. 19 nattanan726/Shutterstock.com; p. 21 Sean Pavone/Shutterstock.com; p. 23 Gladskikh Tatiana/Shutterstock.com.

Library of Congress Cataloging-in-Publication Data

Names: Pang, Ursula, author.
Title: Trains / Ursula Pang.
Description: Buffalo : PowerKids Press, (2025) | Series: On the move
Identifiers: LCCN 2023051600 (print) | LCCN 2023051601 (ebook) | ISBN 9781499444698 (library binding) | ISBN 9781499444681 (paperback) | ISBN 9781499444704 (ebook)
Subjects: LCSH: Railroad trains--Juvenile literature. | Railroad trains--Parts--Juvenile literature. | CYAC: Railroad trains. | LCGFT: Instructional and educational works.
Classification: LCC TF148 .P35 2025 (print) | LCC TF148 (ebook) | DDC 625.2--dc23/eng/20231120
LC record available at https://lccn.loc.gov/2023051600
LC ebook record available at https://lccn.loc.gov/2023051601

Manufactured in the United States of America

Some of the images in this book illustrate individuals who are models. The depictions do not imply actual situations or events.

CPSIA Compliance Information: Batch #CSPK25. For further information contact Rosen Publishing at 1-800-237-9932.

Find us on